Cocurricular Activities:
Their Values and Benefits

Academic Societies and Competitions
Striving for Excellence

Career Preparation Clubs
Goal Oriented

Community Service
Lending a Hand

Foreign Language Clubs
Discovering Other Cultures

Hobby Clubs
Sharing Your Interests

Intramural Sports
Joining the Team

School Publications
Adventures in Media

Science and Technology Clubs
Ideas and Inventions

Student Government and Class Activities
Leaders of Tomorrow

Theater, Speech, and Dance
Expressing Your Talents

Vocal and Instrumental Groups
Making Music

Science and Technology Clubs
Ideas and Inventions

70550   Science and technology clubs . . .

Mark Haverstock

Mason Crest Publishers
Philadelphia

Mason Crest Publishers, Inc.
370 Reed Road
Broomall, PA 19008
(866) MCP-BOOK (toll free)
www.masoncrest.com

First printing

1 2 3 4 5 6 7 8 9 10

Library of Congress Cataloging-in-Publication Data

Haverstock, Mark.
  Science and technology clubs: ideas and inventions/by Mark Haverstock.
      p. cm. (Cocurricular activities)
  Includes index.
  ISBN 1-59084-897-7
1. Science—Study and teaching (Secondary) 2. Technology—Study and teaching (Secondary) I. Title. Q181.H37554 2005
  507'.1'2—dc22
      2004015859

Produced by
Choptank Syndicate, Inc. and Chestnut Productions, L.L.C.
260 Upper Moss Hill Road
Russell, Massachusetts 01071

Project Editors Norman Macht and Mary Hull
Design and Production Lisa Hochstein
Picture Research Mary Hull

OPPOSITE TITLE PAGE

Participation in high school activities is often a predictor of later success in college, career, and life. If you have an interest in science or technology, getting involved in related activities will create new opportunities for you and give you practical experience beyond that learned in the classroom.

Table of Contents

Introduction

COCURRICULAR ACTIVITIES BUILD CHARACTER

Sharon L. Ransom
Chief Officer of the Office of Standards-Based Instruction
for Chicago Public Schools

Cocurricular activities provide an assortment of athletic, musical, cultural, dramatic, club, and service activities. They provide opportunities based on different talents and interests for students to find their niche while developing character. Character is who we really are. It's what we say and how we say it, what we think, what we value, and how we conduct ourselves in difficult situations. It is character that often determines our success in life and cocurricular activities play a significant role in the development of character in young men and women.

Cocurricular programs and activities provide opportunities to channel the interests and talents of students into positive efforts for the betterment of themselves and the community as a whole. Students who participate in cocurricular activities are often expected to follow certain rules and regulations that prepare them for challenges as well as opportunities later in life.

Many qualities that build character are often taught and nurtured through participation in cocurricular activities. A student learns to make commitments and stick with them through victories and losses as well as achievements and disappointments. They can also learn to build relationships and work collaboratively with others, set goals, and follow the principles and rules of the discipline, club, activity, or sport in which they participate.

Students who are active in cocurricular activities are often successful in school because the traits and behaviors they learn outside of the classroom are important in acquiring and maintaining their academic success. Students become committed to their studies and set academic goals that lead them to triumph. When they relate behaviors, such as following rules or directions or teaming with others, to the classroom, this can result in improved academic achievement.

Students who participate in cocurricular activities and acquire these character-rich behaviors and traits are not likely to be involved in negative behaviors. Peer pressure and negative influences are not as strong for these students, and they are not likely to be involved with drugs, alcohol, or tobacco use. They also attend school more regularly and are less likely to drop out of school.

Students involved in cocurricular activities often are coached or mentored by successful and ethical adults of good and strong character who serve as role models and assist students in setting their goals for the future. These students are also more likely to graduate from high school and go on to college because of their involvement in co-curricular activities.

In this series you will come to realize the many benefits of cocurricular activities. These activities bring success and benefits to individual students, the school, and the community.

Students hold their trophy and celebrate their win at the 2003 FIRST Robotics Competition Championship event in Houston, Texas, which featured a robotics challenge called "Stack Attack."

1

Kicking 'Bot

Move, stack, and climb . . .

It's easy if you're a human—tougher if you're a robot. But the Chaney High School Robotics Team of Youngstown, Ohio, was up for the challenge.

The 2003 FIRST (For Inspiration and Recognition of Science and Technology) Robotics Competition challenge, named Stack Attack, required teams like theirs to design robots to stack a column of crates, safely transport them over a bridge, then return to the top of the bridge. Robots had to fight other robots trying to bulldoze crates back into their own scoring zone. If that wasn't challenging enough, they also had to catch up with one of the other robots and ride piggy-back for bonus points.

The competition was fierce. Many times, the way things worked on paper wasn't the way they worked in the real world of robotics competition. With their eyes focused on the prize, the Chaney Cowboys team took charge and made gutsy decisions to fine tune their 'bot, nicknamed Mad

Sign Me Up

How do you get involved in science and technology clubs and activities? The first place to check is with your subject teachers—math, science, industrial technology, or computers. If there's a program in your school, they probably know about it. Pay attention to school bulletin boards and announcements for activities, especially in the fall. If you're home schooled, check with local home school groups and associations for information on contests in which you can participate.

If there aren't any technology clubs or activities at your school, start your own. Discuss the idea with your principal and your teachers and get their support. Then decide on a good time and place to meet. Put up posters and design flyers to pass out to your classmates, or try publicizing your meeting on school announcements.

These clubs and competitions give you a chance to practice and apply some of that knowledge you've learned in class in a practical way now, instead of waiting years. A school without cocurricular activities is a bit like having sports practice without games. "How much enthusiasm and dedication would you get from high school football players if you told them, we will practice very hard each and every day after school, but we will never have a football game?" says Nicholas Frankovits, a high school science teacher from Akron, Ohio. He says that students need games to play that use the skills they learn in class, and these competitions are the games.

All the robotics, math, science, and invention competitions that follow in this book can help make you smarter and your school experiences more fun. Having these activities on your record will also make you look good when it comes to applying to colleges and technical schools. Your participation in cocurricular activities such as these lets them know you're a well-rounded person with many interests.

Get involved, and exercise your mind as well as your body.

Cow, as the contest conditions changed. Their strategy paid off in the end.

"Mad Cow is pushed down the ramp by the Bobcat," said the announcer. "Mad Cow doesn't give up and moves some crates into the scoring zone.

"Mad Cow's still putting things in their scoring zone. Bobcat's still trying to hold Mad Cow at the ramp," continued the announcer. "Mad Cow's at the top of the ramp. There goes the battle with ten seconds left. Oh man! Mad Cow's the only robot today to climb up and hold onto another robot!"

Mad Cow and the Chaney team got fifty bonus points, mainly because of their unusual drive system. They really did reinvent the wheel. "The design was based on a hay baler," says Rick Willmitch, a Chaney team advisor. "It basically walks like it has snowshoes. It's not driving across the

For the two-minute Stack Attack competition, teams designed robots that could collect and stack plastic storage containers on the side of a playing field and keep opposing team robots from doing the same activity.

ground—it's actually lifting and moving forward." During their tests, Mad Cow walked across six inches of snow and climbed up a tall pile of snow. It will even climb steps, something that's unheard of for even the most sophisticated robots.

The Mad Cow's unique way of walking has sparked interest in firefighting, police, and military applications. "Since then, we got a patent on the design and we have some military contractors interested in the design," says Willmitch. "This is a first for FIRST—a new technology being patented and marketed from a design that came from one of their contests."

Fast Facts: Robots

Lend a Hand. Though the robots of movies and TV, like Data of Star Trek, resemble humans, most in the real world don't. Typically, robots are multi-jointed arms or hands that work on assembly lines, putting together a variety of products.

Ring a Bell? One ancestor of the modern robot was the automaton. In the fifteenth century, Europe was filled with church clocks featuring human-like automatons with moving arms that lifted hammers and struck bells.

Super Sized. When electronic computers first came out in the 1940s, they were way too big to fit on a robot. Even the smallest ones would fill rooms—even small buildings. It wasn't until the invention of the microchip that computers were tiny enough to go along for the ride.

Man's Best Friend. Is robo-dog the pet of the future? Several years ago, Sony created Aibo, one of the most advanced "toy" robots yet. The current version of this cyber-pooch recognizes more than one hundred voice commands and can even express emotions like happiness, sadness, fear, dislike, surprise, and anger.

The military would like to use the Mad Cow robot as a minesweeper or to help rescue wounded soldiers in the field. Willmitch says they can easily add a camera to see where it's going. "Soldiers can remotely drive the robot while watching video from it on the computer screen." The team even sent a video presentation of their robot to Iraq, where several army officials got a first peek.

Firefighters might use their robot to find people trapped in a burning building. Police could use it to help sniff out bombs or be their eyes and ears in dangerous situations. That's really impressive stuff coming from a bunch of high school students.

Robotics activities, such as the FIRST competition, are one of several science and technology programs in which you can participate through your school. "Join—it's a lot of fun," says Mike Willmitch, a junior class team member who started back in eighth grade in the FIRST LEGO® League competition. Mike was always handy with tools, but he learned a whole world of other skills in his high school science and technology activities, including computer programming and welding.

FIRST competitors work at the remote controls for a robot. FIRST pairs students with adult mentors who help them prepare for competition.

2

Robotics

The term robot comes from a Czech word meaning "drudgery" or "forced labor." Robots were originally designed to do jobs that were too boring, difficult, or dangerous for humans. They started out as simple machines in ancient Greece and evolved into the computer-controlled mobile marvels we see today.

But the robots built for today's club competitions are hardly drudgery to assemble or boring to build. Ranging from the size of a remote-control car to a small ATV, these smart machines can be trained to follow marked paths, avoid obstacles, and even pick up things.

You don't have to be a rocket scientist or techno-geek to participate in a robotics club. Students often come in with no background. They're just interested in learning how to build a robot. Wendy Wooten, robotics team advisor for Chatsworth High School in Chatsworth, California, says they take "any kid who's enthusiastic, energetic, and wants to get involved in a team sport—this is a sport for the mind.

It's a great opportunity to learn about mechanical things, computers, and electronics."

Chatsworth has a four-year program. "When the freshmen come in, they're the rookies," says Wooten, "and the older kids become their mentors. They also pair up with local engineers, machinists, welders, and people in the manufacturing industry." These students and professionals teach kids a lot of things they'll need to know about building their robots.

Whether you're in middle school or high school, the process will begin long before competition. Your team will probably start planning a year ahead. If you or your team is new to the game, the best thing to do is start visiting competitions and see what other teams are doing. Many of these contests are held in late winter or early spring. Admission is free, and you can learn a lot by talking to the team members. Another way to learn is to go to one of the contest Web sites and read information about past years' competitions.

At the start of the school year in September, you will probably assemble your team, maybe do some fundraising, and start working on some designs. Typically, a team will consist of at least five students, or as many as one hundred in some schools.

Once you get the contest rules and requirements, you'll know what your robot is expected to do. Then you begin building and testing your robot for your first local competition. You'll try some things that work and others that won't. You'll build and rebuild until it comes out right. "The robot becomes like their firstborn child," says Wooten. They named theirs Homer (Human Operated Mechanical Engineered Robot) after *The Simpsons* character.

Just being at a competition with your teammates and

The Right Stuff

What's it like building a winning robot? Boardman Center Middle School Robotics Team Member Dan Bloomberg and his coach, Paula Ritter, share some tips and tricks. The team has chalked up an impressive 3–0 record at the Northeastern Ohio Robotics Competition.

I did what you told me to . . . Robots always do exactly what you say, but not always what you want or expect. "The problem usually isn't with the robot, but the person who is programming it," says Bloomberg. "You have to learn what commands to give it and at what time. Sometimes it's just trial and error until you get it right."

Patience . . . It can take a very long time to make small changes. "Programming the robots is long and repetitive because each time we would change something, it would take about twenty minutes to make the change and another ten minutes to test it," says Bloomberg. "That would be almost an entire morning session just to do one thing."

If at first you don't succeed . . . "We built fifteen different designs before we found one that we liked," says Bloomberg, "and they were all evolutionary designs. We'd begin with one design and add to it until it became something completely different." They finally built one that wouldn't break, was usually reliable, and maneuvered easily.

Teamwork . . . "The biggest factor for a winning group is teamwork," says Ritter. "And by far that's the biggest challenge. You must learn to work with other people who have different personalities."

Think outside the box . . . "Sometimes the direct way to something is not the best way," says Ritter. "Sometimes it's a matter of changing the gear ratios or going about the problem in a different way. Instead of having it just turn left, you might have it go further and make a bigger turn. It's best if you're willing to think in different ways and use different approaches."

being excited about your matches is similar to the thrill you get from sports. "Anticipation, thinking about how you match up against your opponents, and just that whole competitive spirit gets the kids excited," says Wooten. "If you're at a FIRST competition, you've got kids yelling, screaming, jumping up and down, rooting for their team. If you didn't see that there were robots out on the floor, you'd swear you were at a football or basketball game."

There are other opportunities on the sidelines as well. "We get kids who are very artistic but not mechanically minded," says Wooten. "They do things like working on our team logo or designing our newsletters, team uniforms, and T-shirts." Teams also need members who can handle public relations and help raise money to buy uniforms and supplies to design and fabricate the robots, and to travel to competitions all over the United States.

Maybe you're not going to be the quarterback on the football team, or the star center on the basketball team, but there's something for everybody in robotics. Here are a few of the many robotics contests you can enter.

FIRST ROBOTICS COMPETITION
<www.usfirst.org>

FIRST (For Inspiration and Recognition of Science and Technology) is the largest high school robotics contest with over nine hundred teams from around the world participating in robotic design. The competition challenges teams of students and adult mentors to assemble robots using a standard kit of parts (KOP) to do specific tasks that change each year of the contest. Robots can weigh a maximum of 130 pounds, be 30 inches wide, 36 inches long, and 60 inches high. All the building and testing must be done within six weeks.

'Bot Builder to Engineer

Tim Dresser, an engineer at BAE Systems, started building robots in high school for FIRST competitions. "I believe contests like FIRST are fun and interesting—you end up learning more than you think," he says. His experience with a FIRST team taught him what engineers do and gave him valuable experience working on team projects.

According to Dresser, it's not just about the competition. "There are so many different aspects to FIRST. It's not just build a robot and compete," he says. There's a whole lot more involved. You learn to plan, work with deadlines, and solve problems.

You also get to work with the experts. "Our school advisor was trained as an electrical engineer, and the company that supported us sent mentors who were also engineers," he says. "It was really cool. In my senior year, we got to spend two or three afternoons at the company site where we wrapped up the robot assembly and did some test runs." They also got an opportunity to see where their mentors worked and some of the other projects they were working on, including other robots.

FIRST competitions are different from sports-based competitions. You don't want to just dominate and destroy. The object is to help and support your opponents during the competition, but beat them in the end by coming out with the better machine or the better performance.

In many ways, it's a lot like the real world of engineering. "The ability to work with other people is an important skill," says Dresser. "You have to realize that someone who is your competitor today might be your teammate next week. For example, there are projects in which we compete against Boeing, and there are also projects where we team up with Boeing."

Dresser is still involved with FIRST. He volunteered at last year's regional contest and will again this year. His company has sponsored the New Hampshire regional contests since 2003.

Grants, Contests, and Competitions–Robotics

Botball
KISS Institute for Practical Robotics
1818 W. Lindsey, Bldg. D, Suite 100, Norman, OK 73069
ph: (405) 579-4609 **email:** botball@kipr.org
www.botball.org

FIRST
FIRST Robotics Competition/FIRST LEGO® League
200 Bedford St., Manchester, NH 03101
ph: (603) 666-3906 **email:** info@usfirst.org
www.usfirst.org   www.firstlegoleague.org

Ohio TECH Robotics and Technology Invitational
OhioTECH
River Valley Middle School, 4334 Marion-Mt. Gilead Rd.
Marion, OH 43314
ph: (740) 725-5799 fax **email:** tad_d@treca.org
www.ohiotech.org

MATE ROV
MATE Center
Monterey Peninsula College
980 Fremont Street, Monterey, CA 93940
ph: (831) 645-1393 **email:** info@marintech.org
www.marinetech.org/rov_competition/index.php

Once the robots are completed, teams can enter their robots in one of twenty-six regional competitions. Winners attend a Championship event where more than seven thousand students participate. Teams come from all over the world.

Winning a competition doesn't mean scoring a ton of points in head-to-head competition. FIRST teams are rewarded for excellence in design, team spirit, professional behavior, maturity, and the ability to overcome obstacles. It's normal to see team members helping other team members with parts or advice.

Dean Kamen, inventor and founder of FIRST, says it's all about teamwork. "The secret of building a successful team is not to assemble the largest team possible, but to assemble a team that can work well together." FIRST teams have an average of about twenty-five members.

FIRST LEGO LEAGUE COMPETITION
<www.firstlegoleague.org>

FIRST LEGO League (FLL) competition is the little league of the FIRST competitions. It was started in 1998 when the LEGO Company partnered with FIRST to create a program for kids ages nine to fourteen. In 2003, more than 42,000 kids from 48 states and 14 countries participated, and the number continues to grow every year.

Like the FIRST Robotics Competition, FLL isn't about getting the highest score in a tournament. It's about getting experience in creative thinking, figuring out how to solve problems, and learning to work together with other team members.

You build your robot with the LEGO MINDSTORMS™ Robotics Invention System. Using LEGO bricks, motors, sensors, and gears, you get hands-on experience constructing a robot that's about the size of a small radio-controlled car. You'll also learn some computer programming using an icon-based programming language you drag and drop on the screen. Pick the icon that represents the function you want it to do, like turn the motor on or off. Link a string of these icons together, and you have a routine for your robot to follow. The task challenges for the robots change each year.

Besides building and programming robots, FLL team members take part in research assignments, keep journals, and make presentations to show what they've learned at the competitions. These presentations can be anything from skits to cheers, short videos to singing.

What's it like in the competition? "It's a cross between nervous and so excited–you feel like you're going to pass out. It's like a sport because you get psyched up about it, but you're not running around and getting sweaty," say FLL team members Josh Perry and Brian Spradin.

BOTBALL COMPETITION
<www.botball.org>

Botball is played on a four-foot by eight-foot board where robots score points by placing white or black balls in scoring position. The game board setup and scoring rules change each year. Botball requires robots with brains. These robots are autonomous and are trained by students before the competition, using laptop computers and a programming language called Interactive C.

Kids have six weeks to program, design, and build their robots. As they build, they sharpen their science, computer programming, and engineering skills—while having lots of fun.

The robot kit includes microprocessors, sensors, motors, gears, and software—as well as an assortment of LEGOs. They have the option of building two robots that can work as a team during the games.

During the six-week work period, students also partici-pate in a Web site competition where they do research and create a Web site presenting their solution to the year's challenge assignment. One year's challenge was "Design a robot to visit an Apollo landing site." These entries are judged on originality and technical skills.

At the end of the six weeks, teams bring their robots to a regional tournament where they compete. Robots run unopposed in the initial rounds and then compete in double elimination matches. Awards are given for top 'bot, best seeding score, best Web site, and a variety of other criteria. Regional winners are invited to the national Botball tournament.

OHIO TECH ROBOTICS AND TECHNOLOGY INVITATIONAL
<www.ohiotech.org>

The former Robotic Technology and Engineering Challenge (RTEC) has now become the national OhioTECH competition. OhioTECH sponsors contests open to middle school, high school, and college students anywhere in the United States. Students are judged on how they apply technology and their ability to solve real-world problems as a team.

The Olympic-style competition held each April has a number of fun contest categories. These include sumo robots that try to push each other outside a circle, navigate mazes, gather objects on a simulated moonscape, and attempt other tests of robotic skills.

Unlike the other robotics competitions, OhioTECH includes a total of ten separate contest categories, each with its own design requirements.

MATE ROV
<www.marinetech.org/rov_competition>

Ever heat of robots that swim? The Marine Advanced Technology Education Center Remote Operated Vehicle Contest (MATE ROV) holds underwater robotic competitions open to grades six through twelve. The contests are held at regional locations including California, New England, Texas, and Hawaii.

The goals of the MATE ROV competitions are to increase student awareness of marine technical fields and to encourage and prepare students to participate in them. The competition also fosters connections between students and educators and professionals working in the marine field. Businesses, research institutions, government agencies, professional societies, and foundations all partner in the program by donating funds, equipment, supplies, facilities, technical expertise, and their time.

Building underwater robots is quite different—think of a boat vs. a submarine. All the electronics have to be waterproofed to keep them dry. These robots use tethers, wires that supply power and help control the robots. Sometimes it's hard to see what's going on under water, so team members have to use a video monitor to guide their robot.

Home school teammates Sarah Thain, left, and Beckie-Anne Thain, right, test out their ROV, the Nina Harper, in the swimming pool at their family home in British Columbia, Canada. The team tied for first place in the 12 volt 25 amp class of the 2003 MATE National ROV Competition, held at the Massachusetts Institute of Technology.

Shown is the ROV of the Brownsville, Pennsylvania High School Team, who received technical and financial support from Pennsylvania State University's Applied Research Laboratory. Their ROV is about to recover precious "jewels" at the 2002 MATE National ROV Competition held at the Kennedy Space Center and Brevard Community College.

There are two competition classes, General for beginners and Ranger for returning teams. Both involve pool events with the robots and a team display. Ranger class also includes an engineering evaluation where a panel of judges will ask you questions about your robotic project. Some regional competitions, like the one held at the Monterey Bay Aquarium in California, limit the number of teams that can participate.

One thing that's different about this competition is that there is no entry fee, and teams who participate get funds from the contest. If you're a regional winner in the Ranger competition, they'll also help pay for your travel, room, and board while you're at the national competition.

Math skills are important in everyday life and are used in everything from decision-making to home building and decorating, cooking, shopping, and financial planning.

3

By the Numbers

Since the beginning of mankind, humans have been using their fingers and toes to count. They needed to keep track of their herds of animals or the days since the last full moon. To represent and calculate these amounts, they also used things like pebbles, marks on wood, or knots in a rope. By about 3000 B.C. the Egyptians were using a decimal system, counting in groups of ten—the same counting system we use today.

Today you can use pencil and paper or a calculator to help you develop your math talent. Once you've conquered the basics, you may be interested in learning about more advanced mathematics.

Participating in math clubs and contests takes some mathematical skills, but regular practice is even more important. "I start in October and work two mornings a week with students until the contest in February," says Lynn Flaviano, a MATHCOUNTS coach at Boardman Center Middle School in Boardman, Ohio. Though all students are

Every culture on earth has developed some form of mathematics. Roman numerals were once widely used but were eventually surpassed by Arabic numbers, the ones we use today. One way to keep track of numbers in the ancient world was on an abacus, or counting board.

welcome, many of the kids on her MATHCOUNTS team are in the accelerated math program.

During their twice weekly practices, they learn a lot about the different ways to solve problems. "I'm very formula-oriented, while they're very practical in problem solving," Flaviano explains. "They'll come up with solutions in very different ways than I'd expect." She says it's good to share these diverse ideas because it teaches them different ways of thinking and problem solving.

Once it gets to crunch time, students get practice tests for all rounds of the competition. To get her final four qualifiers and alternates, she keeps track of student points along the way. In January, she picks her MATHCOUNTS team based on their point totals.

What's it like in the contest? Her "mathletes" compete

individually in the first two rounds. Then they prepare to face the team round—the most important one. "You can use a calculator, but you really need to develop a strategy to be successful," says Flaviano. "We usually split up the problems, giving the tougher ones to the better students. If everybody worked on the first problem, they'd never get done in time."

Winning is based on points. They total the points from the first three rounds, and the team round counts double. Then they come up with an average per team for the thirty-four teams in their regional contest. "This year's winners had 24—we were in fourth place with 23.25," says Flaviano. The top teams are often close, separated by only hundredths of a percentage point.

"Some kids feel intimidated by the top scorers and don't want to compete against them," says Maggie Wellington, another coach at Boardman Center Middle School. "I think if they would just take a few cards a night and work with them, they'd pick up the knack for finding these combinations. Then they could become some of the top scorers."

It's not always the smart kids who win math competitions. Students who are willing to work hard have a good chance of making the finals and maybe even placing. Think of math competitions in the same way you'd think of trying out for a sport. Although most middle school and high school athletes realize they won't make a living playing professional sports, they still enjoy being involved and challenging themselves to do their best.

Participating may also help your understanding of math. It's also a big help for students who want to get ahead. "The extra practice has helped me through math," says Greg Knight, an eighth grader. "I feel like I know things before others in the class know them."

The odds are in your favor that there's a math club or contest out there for you.

MATH OLYMPIADS
<www.moems.org>

Math Olympiads sponsors two divisions, one for students in grades four to six and another for grades five through eight. Teams meet weekly for an hour, practicing problems or exploring a math topic or strategy in depth.

"I had always said, 'I can't do it' and now I don't say that anymore," says Lindsay McGrory, a fifth grader. "Math Olympiad has given me challenges and I enjoy it."

Each team has up to thirty-five students, and schools can have more than one team. Contestants participate in five monthly contests given from November through March. Each monthly contest given in participating schools has five problems worth one point each. Team scores are based on the sum of the ten highest individual scores, taken after the fifth contest.

Participants and teams receive trophies, medals, pins, and Olympic patches based on their performance.

AMERICAN MATH COMPETITIONS
<www.unl.edu/amc>

The AMC 10 includes math topics usually covered in grades nine and ten. AMC 12 covers the entire high school math curriculum, except calculus. Both are pencil and paper tests that are given in your school and take about seventy-five minutes.

Students who score in the top 5 percent of either test or score at least one hundred points on the AMC 12 are invited to participate in the American Invitational Mathematics Examination (AIME) in March. The top scorers in the AIME are eligible for $1,000 scholarships and are invited to participate in the USA Mathematical Olympiad (USAMO) held in May. The top three USAMO winners can qualify for scholarships ranging from $5,000 to $15,000.

MATHCOUNTS
<www.mathcounts.org>

MATHCOUNTS is a nationwide competition for middle school students in grades seven and eight. About 35,000 "mathletes" participate in one of 500 local competitions.

Grants, Contests, and Competitions–Math

24 Game
Suntex International Inc.
3311 Fox Hill Road , Easton PA, 18045
email: info@24game.com
www.24game.com

American Mathematics Competitions
University of Nebraska, Lincoln, NE 68588
ph: (402) 472-2557 email: amcinfo@unl.edu
www.unl.edu/amc

MATHCOUNTS
MATHCOUNTS Foundation
1420 King Street, Alexandria, VA 22314
ph: (703) 684-2828 email: info@mathcounts.org
www.mathcounts.org

Math League
Math League Press
P.O. Box 17, Tenafly, NJ 07670
email: comments@mathleague.com
www.mathleague.com

Math Olympiad
2154 Bellmore Ave.
Bellmore, NY 11710
ph: (866) 781-2411 email: info@moems.org
www.moems.org

Top teams and top individual scorers are chosen for the state competitions, and the four winning individuals at the state level advance to the national competition in Washington, D.C.

In MATHCOUNTS there are four rounds; the first is called the Sprint Round. Each contestant does thirty problems individually, without the help of other team members or calculators. Next comes the Target Round, where students answer eight questions, given to them two at a time with six minutes to answer each pair. The Team Round consists of ten problems in which all team members work together. Finally, the Countdown Round pits individual students in head-to-head competition.

Go Figure

Contests like MATHCOUNTS aren't just about crunching the numbers. The afternoon is filled with fun activities. "We clap, yell, scream—really get pumped up," says Tyler Reed, a member of the Canfield Middle School Team in Canfield, Ohio.

First it's the Math Bowl, where school teams of four face three other school teams. You start with two hundred points and wager them Jeopardy-style after seeing the question. You get it right, you keep the points; get it wrong and the points are gone. "We usually bet it all every time," says Reed. "You don't win unless you take a chance." The team with the most points after four questions moves to the next round. The battle continues until only two schools are left to fight for the title.

The Countdown Round takes the top sixteen people from the morning contest, who are matched in head-to-head competition. With hands on the button, the first to buzz in and give the correct answer to the question moves on. When the contest is down to four, it's the best out of three questions until a winner finally emerges.

Mathematics is the science of numbers. Math teaches thinking and problem solving, and math skills have many applications in everyday life.

MATH LEAGUE CONTESTS
<www.mathleague.com>

The Math League sponsors several contests: one for grades four and five, another for grades six through eight, as well as an algebra 1 and high school contest. More than one million students from the United States and Canada participate in Math League contests every year. School champions compete in statewide or multi-state league championships.

Every contest has questions from a variety of areas in mathematics. The goal of the contest is to encourage interest in math and help students gain confidence in math by solving problems. Contests consist of thirty-minute sessions of multiple choice questions done with pencil and paper. Calculators are allowed. There are thirty questions for grades four through five and algebra 1; forty questions for grades six through eight. The high school contest also lasts thirty minutes, but consists of six very challenging questions, which increase in difficulty.

24 CHALLENGE MATH
<www.24game.com>

Imagine going into a math competition and knowing the answers to all the questions. In 24 Challenge Math

---

Beating the Odds

Name: Kiley Thompson, 13
Game: 24
Claim to Fame: School 24 Game champion
Thrill of Victory: I made it to both the regional and state contests. When I got to state, I saw my old fourth-grade rival. We were in the semifinals and she got two penalties and was out of the game. I made it to the final four.
Strategy: Practice. You need to know your multiplication, division, addition, and subtraction facts. You do everything mentally—don't even think about paper and pencil.

Name: Greg Knight, 14
Game: MATHCOUNTS
Claim to Fame: Regional champion
Thrill of victory: Going to state competition.
Strategy: I just try to relax before the contest. I think about what I'm going to do and just go do it. I always work on knowing the formulas, especially area and volume.

Name: Tyler Reed, 14
Game: Ohio Math League
Claim to Fame: Eighth place in Ohio, first place in Mahoning and Trumbull Counties
Thrill of victory: I was first in my school by five points and got a lot of compliments from the older kids in the competition.
Strategy: I do a lot of practice tests.

everyone knows the answer is twenty-four, no matter what the problem. The object is to make twenty-four from the four numbers on the game card using addition, subtraction, multiplication, or division. "Our goal is to use the innovative 24 Game as a tool to make math more appealing, accessible, and fun, thereby helping students to move on to excel in math, science, and related fields," says Robert Sun, inventor of the 24 Game.

Cards have three different levels of difficulty, shown by one, two, or three dots on the card face. There's often more than one answer for each card. Winning means knowing how you find the answer or the method behind the math.

School teams usually begin practice sessions in September with students playing against one another. Students from first to eighth grade can participate, but usually the older students participate in the playoffs, which are held in each school to determine who will go to regional competitions. Winners from the regional competitions may participate in state contests.

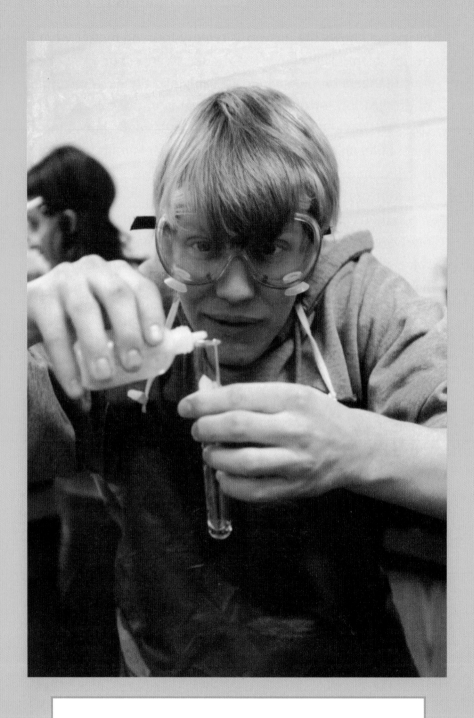

Science fairs give you the opportunity to showcase your research and compete for prizes, including college scholarships.

4

Scientists at Work

Did you ever get the urge to make a scientific discovery of your own? Maybe something that could change the world? Something cool and high-tech? Reach beyond the limits of your science classroom and get some extra credit by experiencing real science in action.

If your school has a science fair or a science club, joining them will help you learn more about science and problem solving. These activities could also be stepping stones to regional and national science fairs and competitions where you can present your ideas to real scientists and researchers, and maybe earn some money for college.

Many companies sponsor science competitions because they want to get more students interested in science and technology. "If we're going to continue to grow, we're going to need professional engineers, researchers, and people who know how to grow technology-based businesses," says Marie Gentile, Manager of Educational Outreach for the Siemens Westinghouse Competition for math, science, and

Grants, Contests, and Competitions–Science

Christopher Columbus Awards
105 Terry Drive, Suite 120, Newton, PA 18940
ph: (800)291-6020 email: success@edumedia.com
www.christophercolumbusawards.com

Discovery Channel Young Scientist Challenge
email: dysc@discoveryschool.com
http://school.discovery.com/sciencefaircentral

Intel International Science and Engineering Fair
Intel Science Talent Search
1719 N Street, NW, Washington, D.C. 20036
ph: (202) 785-2255 email: sciedu@sciserv.org
www.sciserv.org/isef/ www.sciserv.org/sts/

Internet Science and Technology Fair
University of Central Florida
4000 Central Florida Blvd., Orlando, FL 32816
email: director@istf.ucf.edu
www.istf.ucf.edu

Siemens Westinghouse Competition: Math, Science, Technology
Siemens Foundation
170 Wood Avenue South, Iselin, NJ 08830
ph: (877) 822-5233 email: foundation@sc.siemens.com
www.siemens-foundation.org/science/default.html

Science Olympiad
5955 Little Pine Lane, Rochester Hills, MI 48306
email: soinc@soinc.org
www.soinc.org

technology. "We also realize that these are students who work very hard and there aren't as many scholarship opportunities for math and science students as there are for sports." This is an opportunity for students who excel in math and science to also get recognition for their work.

Gentile says there are many beneficial reasons why students, especially ones in high school, enter science and technology contests. "Many do it for the scholarship money, because these are students who will go on to schools like MIT, Harvard, Princeton, Yale—and those schools cost a lot of money. Programs like ours can provide scholarship money." Others just do it because they like science and math.

College admissions counselors are definitely looking at your application for activities like these, because college admissions aren't based strictly on your SAT scores or academic performance. They want to know about you as a person and your interests and accomplishments beyond the classroom.

Preparing for the competition itself can do a lot for you. You get to practice real-world science. "Kids say they like the opportunity to present their work like real scientists would do," says Gentile. "In our competition we require participants to do original research. If they're selected as regional finalists, they come and present their findings to a panel of judges. That's what a real scientist does when he stands in front of his colleagues at a symposium."

When you participate in a science contest, you learn about time management skills, finding a project, researching it, and following it through to completion before your deadline. You might even get a chance to work with real scientists or attend camps to help you on your way.

Yuyin Chen, a seventeen-year-old student at Cranbrook Kingswood School in Bloomfield Hills, Michigan, followed his interests in math and entered the Intel Science Talent search. "I heard about the Intel competition at a science research camp held at MIT called the Research Science Institute," he says. "We were matched with mentors from the Institute, and mine was a graduate student in MIT's math program." When the camp was over, his mentor encouraged him to submit his project, a type of cutting problem in graph theory. "Cutting problems have a lot of applications in computer and communication networks, called very large scale integration systems (VLSI)," says Chen. "One practical example of VLSI is the computer microchip."

His advice is to find a project you really like and work hard on it. "You're going to spend a lot of time researching and doing background work." Ideas? Think of things you've done in the past that have interested you. Chen's project was based on his interest in graph theory that started back in his freshman year.

Fast Facts: Preparing the Way

If you want to do a really good job on your project, there are some basics you need to know.

1) Understand the Scientific Method. It's important to understand the rules of scientific research before you begin.

2) Choose your topic. Find something that interests you. Remember that science projects are really tests you do to find answers to questions, not just to show what you know. Also think of topics that can benefit your community or solve a problem.

3) Run your controlled experiment and record the results. Be sure to follow the Scientific Method and take good notes; you'll need them later.

4) Compile your research, prepare your presentation and/or build your display. Make it interesting and neat, so it clearly explains the story of your project. Be sure you do these according to contest requirements.

5) Practice your presentation and be prepared to answer questions about your work.

Do you have to be a geek or nerdy type to be successful? "It's certainly not true," says Chen. "There are a lot of people that are good at math or science, and you'd never guess it. For example, our star linebacker tops our school in math competitions—he even beats me." All kinds of people can be good at math or science.

A variety of local and national contests are just waiting to hear about your scientific discoveries.

INTEL SCIENCE AND ENGINEERING FAIR
<www.sciserv.org/isef>

The Intel International Science and Engineering Fair (Intel ISEF) is the world's largest celebration of high school science. The event is held every year in May, and the competition brings together more than 1,200 students from 40 nations to compete for scholarships, tuition grants, internships, and scientific field trips. The grand prize winner receives a $50,000 college scholarship and a high-performance computer. Science Service originally started the ISEF in 1950.

To qualify, you must compete in a local fair that's affiliated with Intel ISEF and be selected to represent that fair. Each year, between 3 and 5 million students complete science fair projects and about 1,200 of these qualify to compete at the Intel ISEF. Two individual winners and one team project from this contest get to compete at the international level.

Students compete for scholarships in fifteen categories: Behavioral and Social Sciences, Biochemistry, Botany, Earth and Space Sciences, Chemistry, Computer Science, Engineering, Physics, Environmental Science, Gerontology, Mathematics, Medicine & Health, Microbiology, Team Projects, and Zoology.

Ryan Patterson, winner of the 2002 Siemens Westinghouse Science and Technology Competition, poses with his invention, a glove that translates sign language into written letters.

INTEL SCIENCE TALENT SEARCH
<www.sciserv.org/sts>

Intel and Science Service also sponsor the Intel Science and Talent search (Intel STS). Their goal is to find and encourage talented high school seniors to choose careers in science, math, engineering, and medicine. Each year, almost two thousand students enter, and forty finalists are chosen to compete for the top prize, a $100,000 scholarship.

Many students, like Yuyin Chen, take part in a science training program (STP) to help them with their project. More than three hundred training programs are held throughout the year at a variety of locations, including U.S. colleges and universities. These programs offer advanced students instruction and hands-on lab research.

Talk to the Hand

Ryan Patterson, a high school student from Grand Junction, Colorado, watched a sign language interpreter help his hearing-impaired companions order a meal at a fast-food restaurant. "It made me realize that there was a need for another kind of interpreter, something portable that could be used when a human interpreter isn't available."

After some brainstorming, he came up with a whopper of an idea. Using an ordinary golf glove, sensors, a wireless transmitter, and a computer processor, Patterson invented a sign language translator to go. Slip on the glove, and any hearing-impaired person can translate the finger movements of American Sign Language (ASL) into letters that appear on a lighted candy-bar-sized display.

Patterson spent almost one thousand hours perfecting his device over a period of eight months. One size doesn't fit all. It needs to be programmed by the individual user to adapt to his or her own style of signing. He plans to make other versions of the talking hand, including one that translates sign language to speech and another that can be used along with personal digital assistants (PDAs).

The sign language translator is a result of his lifelong interest in scientific research, especially in electricity and electronics. According to Patterson, who's also a mountain biker and water sport enthusiast, "Science is fascinating—it's exciting and it's a lot of fun." He suggests that students participate in science competitions. "I'd encourage trying it no matter what. You don't have anything to lose and it can definitely be worth it."

Patterson won first place in the 2002 Siemens Westinghouse Science and Technology Competition and $100,000 in scholarship cash. He also had a chance for some first-hand meetings with famous scientists. Patterson attended the University of Colorado to study electrical engineering, which should help him to develop more cool projects in the future.

Sometimes considered the "junior Nobel Prize," the Intel STS challenges young scientists to look beyond the school classroom and follow their own interests in science and technology. Past finalists in the program hold more than one hundred of the world's highest science and math honors, including five Nobel Prizes.

SIEMENS WESTINGHOUSE COMPETITION
<www.siemens-foundation.org/science/default.html>

The Siemens Westinghouse Competition in Math, Science, and Technology seeks talented high school students who want to challenge themselves through science research. Students who participate have an opportunity for national recognition for their science research projects.

Students begin by submitting research individually, or in teams with two or three members. These are read and judged by scientists from leading universities and labs. The projects that are picked from this group move on to one of six regional competitions. At each of the regional competitions, an individual and team are selected to move to the national competition in Washington, D.C.

This last part of the contest is judged by research scientists who are experts in the subjects covered by the projects in the national competition. Everyone who makes it to the national level receives scholarships of at least $1,000 and the top participants can earn additional scholarships ranging from $10,000 to $100,000.

CHRISTOPHER COLUMBUS AWARDS
<www.christophercolumbusawards.com>

The Christopher Columbus Awards challenge teams of middle school students to use their creativity to make a difference in their communities. Identify a problem, come

In 2001 students at Pretty Eagle Catholic School in St. Xavier, Montana, won an award from the Bayer/National Science Foundation and received a $25,000 Columbus Foundation Community Grant to build a straw bale community center on the Crow Reservation.

up with a solution, and then test it scientifically. Teams of three or four kids prepare entries and show judges how their solutions will work. If your group project makes the cut, you could become one of eight finalist teams to travel to Disney World in Florida, win a $2,000 savings bond, and get a chance for a $25,000 grant

The Columbus Foundation Community Grant gives one finalist team the opportunity to make their project a reality the year after the competition.

Finalist teams also have a chance to attend Christopher Columbus Academy, where students and their team coaches have a one-of-a-kind opportunity to work with engineers,

The Last Straw

For high school students Lucretia Birdinground, Kimberly Deputee, Omney Sees The Ground, and Brenet Stewart, the housing shortage on the Crow Reservation in Montana really hit home. It wasn't unusual to see large families or multiple families crowd into trailers or cheaply-built government housing.

The girls and their teacher, Jack Joyce, began discussing solutions. One affordable idea was to use straw bales as a building material instead of wood or bricks. Straw was plentiful and cheap, with more than 9 million bales available in the county each year. Bales would be stacked around wood framing and then covered with chicken wire and stucco. Topping off the structure would be a metal roof.

Though the girls were sold on the idea, adults were skeptical. Wouldn't straw burn easily or rot from moisture? And of course they had to endure all those lame Three Little Pigs jokes.

"The kids found a problem in their community and used science to solve it," says Joyce. They designed scientific tests to help change these misconceptions and build community support for straw houses. First they made a model of the house wall. They held a blow torch to the side, but it didn't burn. To test energy efficiency, they held thermometers on the inside of the bale while they torched the other side, but the heat wouldn't travel through it. They stuffed rags inside the bales and squirted a hose on it for twelve hours, but the rags remained completely dry.

Their experiments proved to the Crow Nation that straw bales were a safe, efficient building material, and their idea became an award winner. The girls won a Bayer/National Science foundation award, $25,000 from Westmoreland Resources, a $25,000 Columbus Foundation Community Grant, and $25,000 from Oprah Winfrey. All the prize money was used to build a straw-bale community learning center. A small army of volunteers, including the four girls, built it in just two and one-half days in July 2002.

Team members compete at the 2003 Science Olympiad, one of the premier science competitions in the nation. The teamwork required in many Science Olympiad events parallels the group dynamic of scientists in real-world settings.

scientists, and technical gurus while checking out all the fun science behind Walt Disney World.

DISCOVERY CHANNEL YOUNG SCIENTIST CHALLENGE
<http://schooldiscovery.com/sciencefaircentral>

The Discovery Channel Young Scientist Challenge (DCYSC) is open to middle school students. The first step is to present your project at a local or regional International Science and Engineering (ISEF) fair. Every year, student entries from these fairs are nominated to enter their projects in the DSYSC. Of these, four hundred are chosen as

Students at Indiana's Grayslake Middle School get so fired up about participating in the Science Olympiad, they hold a pep rally prior to the event. Nearly 5,500 middle and high schools participate in Science Olympiad each year.

semifinalists. Projects are judged on their scientific merit and on the entrant's ability to explain the science behind the project.

Forty finalists get a trip to Washington, D.C. for a week in October. Here, they work in teams to take on challenges presented by Steve Jacobs, creator of *Jake's Attic*. Many of these activities are done at the Smithsonian and cover a variety of science topics. Contestants' scores are based on presentations of their original science project entry and their participation in the team challenges. Winners receive scholarships ranging from $500 to $15,000.

INTERNET SCIENCE AND TECHNOLOGY FAIR
<www.istf.ucf.edu>

The Internet Science and Technology Fair (ISTF) is a Web-based science and technology competition. It includes school teams from grades three through eight from more than twelve states and two countries, and it's growing. Enrollment begins in the fall and judging begins in March of each year.

Students work as a team, directed by a teacher, to complete a four-month research project by doing research on the Web. They contact scientists and engineers using e-mail, as well as search and analyze electronic sources on the Internet.

Instead of creating some poster board displays or live presentations, teams present their final research findings in the form of a Web site. These Web pages are judged against all the other entries to choose the award winners. No traveling is involved; everything is done at your school.

SCIENCE OLYMPIAD
<www.soinc.org>

Science Olympiad sponsors intramural, district, regional, state, and national tournaments that bring science to life, show how science works, and emphasize the problem-solving aspects of science. The competitions follow the format of popular board games, TV shows, and sports. Disciplines include biology, earth science, chemistry, physics, computers, and technology, and competitors are tested on their knowledge of science facts, concepts, processes, skills, and applications. One of the goals of Science Olympiad is to elevate science education to a level of enthusiasm and support normally reserved for athletic programs.

Kavita Shukla's science project on the antibacterial properties of the herb fenugreek earned her the 2002 Lemelson-MIT High School Invention Apprenticeship.

5

Invention Connections

Thomas Edison was probably the most gifted inventor of the last century. He had almost eleven hundred patents to his name. These included such familiar inventions as the light bulb and the phonograph. But Edison didn't go it alone. He set up an "invention factory" at Menlo Park, New Jersey, and gathered a team that worked together to come up with ideas and solve problems. Remember the old saying, "Two heads are better than one?"

Some invention clubs and competitions stress the team approach and concentrate on producing inventions that fill a need or solve a problem. "It's not so much about competition. Our goal is to have student teams go through the invention process, learn about invention, and invent something in the end—a product or prototype," says Josh Schuler, a Lemelson-MIT InvenTeams grants officer. "We want students and teachers to really experience hands-on science and math through the invention process, something different from what they do every day in the classroom."

Accidents That Became Awesome Inventions

Think you made a mess of your invention? Don't give up. You may find things you were never looking for, things that could change the world, or at the very least, taste really good.

Microwave Oven. In 1946, Dr. Percy LeBaron Spencer was working near radar equipment. He had a candy bar in his pocket, and found that it had melted. After a while, Spencer realized that the microwaves produced by the radar equipment had caused it to melt. After experimenting, he realized that microwaves would cook foods quickly, even faster than regular ovens.

Post-It Notes. Researcher Spencer Silver was working on a new kind of super-strong glue. But his experiments produced only a super-wimpy glue, one that stuck to objects but peeled off easily. A couple of years later a fellow researcher named Arthur Fry used this glue to coat paper, making bookmarks that would stay in place and could be easily removed. Their company, 3M, liked the idea so much they began selling it in 1980. It became one of the most popular office supplies ever.

Popsicle. Eleven-year-old Frank Epperson forgot about the powdered drink he'd mixed on the porch. It sat overnight in the cold, with the stirring stick still in it. The next day, Epperson had a frozen drink on a stick to show his friends. Eighteen years later, in 1923, he started a business selling these frozen treats.

Penicillin. In 1928 scientist Alexander Fleming noticed that mold spores had contaminated some bacteria samples he had left by an open window. Instead of throwing out his ruined experiment, Fleming took a close look and noticed the mold was dissolving the harmful bacteria. That's how we got penicillin, which helps people to recover from infections.

Coca-Cola. In 1886 a pharmacist named John Pemberton was cooking up some medicine in a large brass kettle. When he was done, he figured he had created a cure for people who were tired, nervous, or had sore teeth. Coke didn't make it as medicine, but eventually it became the world's most popular soft drink.

Have you ever used a post-it note? This office supply innovation was created by accident, when a researcher discovered the glue he was making was sticky, but not strong.

How do you come up with original ideas? "One of the easiest things to do is to think about what problems you face and what problems you observe," says Schuler. Start logging these problems—make a list of things that frustrate you. You might think of things that break easily, have parts that get lost, or are just plain annoying to use.

Brian Short, Program Manager of the Craftsman/NSTA Young Inventors Awards, suggests looking close to home. "I'm sure you have chores and your parents have things they need to do around the house," he says. "Look at ways to make those chores and activities easier."

Once you've identified a problem, start thinking of solutions. The first step, brainstorming, helps generate a lot of ideas—some useful, some wild and crazy. But don't toss out those crazy ones. They may turn out to be usable later in some form. Consider improvements on existing inventions; you can always build a better mousetrap.

While testing a radar-related research project, Dr. Percy Spencer discovered that it was able to melt candy bars, pop corn, and cook eggs. Spencer added a metal box to his invention and developed the first microwave oven.

Chandler Macocha was in seventh grade when he observed his neighbor in a wheelchair and asked her about her needs. That's how he came up with his winning idea for the Craftsman/NSTA Young Inventors Award, a wheelchair backpack holder. "It's basically a big lever that extends around the back under the right armrest," says Macocha. "You push it forward and the whole assembly swings around carrying the backpack from the back to the side of the wheelchair."

Making the invention was fun, but not always easy. First he had to find an old wheelchair he could use to test his idea. "I couldn't hire a machinist to make special pieces for the project," he says, "so I had to use regular hardware—screws, nuts, and bolts like you'd get at Home Depot." Another challenge was making it sturdy enough to hold a heavy backpack. Macocha didn't let minor setbacks get to him. "I think that you should keep going if you think your invention could succeed," he says. "Just believe that you

Grants, Contests, and Competitions–Invention

Craftsman/NSTA Young Inventors Awards Program
1840 Wilson Blvd., Arlington, VA 22201
ph: (888) 494-4994 **email:** younginventors@nsta.org
www.nsta.org/programs/craftsman

Invent America
P.O. Box 26065, Alexandria, VA 22313
email: inventamerica@aol.com
www.inventamerica.com

Lemelson-MIT InvenTeams
The Lemelson-MIT Program
Massachusetts Institute of Technology
77 Massachusetts Ave., Room E60-215, Cambridge, MA 02139
ph: (617) 253-3352 **email:** inventeam@mit.edu
http://web.mit.edu/invent/www/InvenTeam/

Toshiba/NSTA ExploraVision Awards
1840 Wilson Blvd., Arlington, VA 22201
ph: (800) 397-5679 **email:** exploravision@nsta.org
www.exploravision.com

can do it." Macocha has filed for a patent on his invention, and he was inducted into the Inventor's Hall of Fame. "It felt pretty good because if you look at the finalists I'm being inducted with, they're in eleventh or twelfth grade. I'm in eighth."

Participating in invention programs and contests gives you a first-hand look at the scientific and engineering fields. It's also a way for kids to look at science in a different way instead of just reading out of the textbook.

LEMELSON-MIT INVENTEAMS GRANTS
<http://web.mit.edu/invent/www/InvenTeam>

InvenTeams isn't a competition, but a way for high school student teams to go through the invention process,

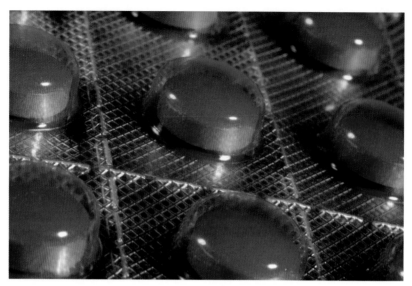

Penicillin, discovered by accident in 1928 when a spore of penicillium mold drifted into the lab of Scottish research scientist Alexander Fleming, is now the most widely used antibiotic in the world. It has become as common as the pills produced on assembly lines and sold in blister packs.

learn about invention, and create a working product or prototype. Schools chosen to participate in the program receive grants of up to $10,000 to help them create inventions that will benefit their school or community. At least ten of these grants are awarded each year.

Learning by doing is the way InvenTeams work. They rely on partners and mentors, such as local businesses and people in technical fields, to help them learn the necessary skills to make their ideas a reality. One group from Smith Academy in Littleton, Massachusetts, developed an energy-efficient underground heating system to clear streets and sidewalks of snow and ice.

During the invention process, teams must send regular progress reports. They show how their prototype works, using video, digital photos, or software. Teams also present their work at the InvenTeams Showcase.

INVENT AMERICA

<www.inventamerica.com>

Invent America helps kids in eighth grade or below to develop creative thinking and problem solving skills through inventing. Entries are judged on usefulness, creativity, illustration, and how well the ideas and research are presented.

Invent America offers recognition and awards from $100 to $1,000 in savings bonds. Past winners have submitted inventions such as the Upsie Daisy, a set of remote control hydraulic jacks installed on an automobile chassis to make tire changing easier.

TOSHIBA/NSTA EXPLORAVISION AWARDS

<www.exploravision.org>

Teams of two to four students from all grades, along with their advisor, choose a technology that's currently in use. It can be simple as a ballpoint pen or complex as a quantum computer. These teams explore what the technology is about

Members of the Arlington (MA) High School InventTeam. In 2004 these students received a Lemelson-MIT InvenTeams grant for their work on an automatic pedestrian crossing device. They were one of only ten high schools in the country chosen to receive the grant, which supports inventions that benefit schools or communities.

A prohibition law led Dr. John Pemberton, a druggist, to substitute sugar for wine in his popular nerve tonic and headache remedy, Coca-Cola, which takes its name from two of the original ingredients, coca leaf and kola nut.

and research the details of its invention. The next step is to imagine what this technology would be like in twenty years and prepare a presentation about it.

Past winners have produced ideas for inventions such as a personal compost/paper consolidator and a land mine detector. Prizes include digital cameras and school laptop computers for regional winners and up to $10,000 in savings bonds for first place teams.

CRAFTSMAN/NSTA YOUNG INVENTORS AWARDS
<www.nsta.org/programs/craftsman>

This contest is strictly for individuals, not teams. Kids can invent something completely new or take an existing tool and modify it in a way that is different or unique. The tool or invention has to perform a practical function, such as mend, make life easier or safer, entertain, or solve an everyday problem. There are two competition levels, grades two to five and grades six to eight.

In addition to making the invention by themselves, contestants need to submit an inventor's log. It can be in one of many different formats, including research paper format, diaries, or question and answer. You must also answer the eleven questions that come with the entry materials. "The more information you can provide the better because it's the only way for you to communicate directly with the judges," says Short.

Prizes range from $10,000 savings bonds for the two national winners to $25 savings bonds for the third place regional winners.

Whether your interests lean towards science, math, computers, or technology, there are opportunities available for you to explore your interest. Consider getting involved in a science or technology activity and put the skills you've learned in the classroom to the test.

Glossary

automaton–a machine or mechanism that can move automatically.

calculus–a branch of mathematics dealing with the way that relations between certain sets (functions) are affected by very small changes in one of their variables (independent variable) as they approach zero. It is used to find slopes of curves, rates of change, and volumes of curved figures.

colleagues–people who are members of your profession; co-workers.

decimal system–the number system we use, having a base of ten, in which numbers are expressed by combinations of the ten digits, 0 to 9.

journal–a daily record of events or observations; a diary.

microprocessor–a computer processor on a microchip; it is the "brain" that goes into action when you turn your computer on.

Nobel Prize–an annual award for outstanding contributions to chemistry, physics, physiology, medicine, literature, economics or peace.

patent–a document granting an inventor sole rights to an invention.

prototype–an original or model from which copies are made.

scientific method–an organized approach to the study of science, involving observation and theory to test scientific hypotheses consisting of the following steps: observation, hypothesis, testing, interpreting results, and conclusion.

symposium–a group meeting where there are speeches and discussions.

very large scale integration (VLSI)–a term describing semiconductor microchips composed of hundreds of thousands of logic elements or memory cells.

Internet

www.robotics.com/robots.html
Robot information central.

http://robotics.jpl.nasa.gov
NASA JPL robotics page.

www.scf.usc.edu/~nmahesh/competitions.html
An archive of math competitions.

www.unl.edu/amc/a-activities/a7-problems/problemdir.html
A math problem directory from the Mathematical
Association of America.

www.ipl.org/div/kidspace/projectguide/
The Internet Public Library science fair project
resource guide.

www.mathforum.org/teachers/mathproject.html
Math ideas for science projects.

www.scifair.org
The ultimate science fair resource.

www.inventorsdigest.com
The online version of *Inventor's Digest*.

Further Reading

Arrick, Roger, and Nancy Stevenson. *Robot Building for Dummies.* New York: John Wiley & Sons, 2003.

Bochinski, Julianne Blair. *The Complete Handbook of Science Fair Projects.* New York: John Wiley & Sons, 1996.

Flatow, Ira. *They All Laughed... From Light Bulbs to Lasers: The Fascinating Stories Behind the Great Inventions That Have Changed Our Lives.* New York: Perennial, 1993.

Hrynkiw, David, and Mark W. Tilven. *Junk Bots, Bug Bots, and Bots on Wheels: Building Simple Robots with Beam Technology.* New York: McGraw-Hill Osborne Media, 2002.

Jones, Charlotte Foltz. *Mistakes That Worked.* New York: Doubleday, 1991.

Pickover, Clifford A. *Wonders of Numbers: Adventures in Mathematics, Mind, and Meaning.* New York: Oxford University Press, 2003.

Rosner, Marc and Scientific American. *The Scientific American Book of Great Science Fair Projects.* New York: John Wiley and Sons, 2000.

St. George, Judith. *So, You Want to Be an Inventor?* New York: Philomel, 2002.

Van Cleave, Janice. *Janice VanCleave's Guide to More of the Best Science Fair Projects.* New York: John Wiley & Sons, 2000.

Wright, Joseph P. and Mario G. Salvadori. *Math Games for Middle School: Challenges and Skill-Builders for Students at Every Level.* Chicago: Chicago Review Press, 1998.

Index

PICTURE CREDITS

Cover: Benjamin Stewart, PhotoDisc.

Interior: © Ron Walloch Photography: 2, 36; Courtesy of FIRST (For Inspiration and Recognition of Science and Technology) Adriana M. Groisman: 8, 11, 14; Courtesy of the MATE Center–Peter Thain: 24; Courtesy of the MATE Center–Steve Van Meter: 25; ©Oscar C. Williams: 26; Photos.com: 28, 33, 53, 54, 56; Courtesy of the Siemens Foundation: 42; Courtesy of the Christopher Columbus Awards/Dymun + Company: 45; ©Jenny Kopach/Science Olympiad: 47, 48; Ken Lam/Courtesy of the Lemelson-MIT Program: 50; Ray Santos/Courtesy of the Lemelson-MIT Program: 57; Lisa Hochstein: 58.

ABOUT THE AUTHOR

Mark Haverstock is a teacher, writer, and technology fan who spends way too much time on his computer. He's written more than four hundred magazine articles for *Boys' Life, Guideposts for Teens, Highlights for Children,* and dozens of other publications. Mark lives in Boardman, Ohio, with his wife Debbie and Cookie the Wonder Dog.

SERIES CONSULTANT

Series Consultant Sharon L. Ransom is Chief Officer of the Office of Standards-Based Instruction for Chicago Public Schools and Lecturer at the University of Illinois at Chicago. She is the founding director of the Achieving High Standards Project: a Standards-Based Comprehensive School Reform project at the University of Illinois at Chicago, and she is the former director of the Partnership READ Project: a Standards Based Change Process. Her work has included school reform issues that center on literacy instruction, as well as developing standards-based curriculum and assessments, improving school leadership, and promoting school, parent, and community partnerships. In 1999, she received the Martin Luther King Outstanding Educator's Award.